From Red Hot Blob to Blue Singing Whale

A Look Into Thermodynamics and the Evolution of Species

Nicosia, 2023

(1st edition)

ISBN: 978-9925-8050-0-6

Introduction

This book offers an engaging exploration of scientific philosophy, mixing established scientific facts and theories with speculative ideas that aim to intrigue the reader and provoke deeper considerations. While I'm neither a scientist nor a philosopher in the classic sense, I believe that all humans are intrinsically both, and this book represents my own exploration of fundamental concepts and their potential links.

The springboard for starting to write this book was my realization that the second law of thermodynamics has been erroneously manipulated or misrepresented to bolster unscientific claims that purportedly undermine the theory of evolution. I felt the need to write my thoughts on the matter, expand on established ideas, and simultaneously rectify false perspectives by illustrating the genuine connection between the thermodynamics and evolution, highlighting the fundamental errors in the reasoning of these misguided convictions while offering a fresh viewpoint to all people regardless of their beliefs.

As sentient beings, we are biologically programmed to seek answers about our identity, a pursuit that invariably draws us into the territories of science and philosophy. Our quest to comprehend the profound depths of these fields pushes us to expand our understanding, frequently to its very boundaries. Exploring our place within the cosmic landscape is an endeavor of profound significance. However, the acceptance of oversimplified, incomplete narratives presented as irrefutable truths is not only misleading but also unethical. These narratives have been birthed to offer solace to those grappling with existential angst, yet their current prominence in the modern world is unjustified.

The main goal of this book is to scrutinize and debunk these distortions while examining the interrelation of thermodynamics and evolutionary theory to offer a solid basis for a new point of view, demystifying the underpinnings of evolutionary theory, presenting scientific facts in a digestible manner, while also introducing an unexpected correlation with the 2nd law of thermodynamics. It aspires to refresh the readers' worldview, enhance their discernment, and guide them in tracing the threads that weave the tapestry of our existence.

This book, however, doesn't entangle the reader in complex scientific terminology; instead, it provides accessible, real-life examples that resonate with the reader's everyday experiences.

Therefore, it caters to those seeking a relaxed exploration of the subject matter, aims to inform, and ultimately, set the readers into a thought-provoking journey into the realms of scientific philosophy.

Chapter One – The 2nd Law of Thermodynamics

<u>Introduction</u>

The second law of thermodynamics often appears enigmatic, leading to confusion and making simple science seem complex. Yet, in essence, its principles are rather straightforward. The ambiguity primarily arises from the fact that the second law of thermodynamics lacks mathematical proof, prompting various analogous explanations and interpretations. Its validity is undeniable, as no counter-evidence has been found, and its principles align seamlessly with observable phenomena in our everyday lives and scientific experiments. Despite this, its articulation and understanding remain a challenge. This chapter endeavors to bridge this gap, elucidating the law, its profound implications for humanity, and its pervasive relevance in our daily lives.

<u>Defining the 2nd Law of Thermodynamics</u>

German theoretical physicist Max Planck defines the second law of thermodynamics as follows: "Every process occurring in nature proceeds in the sense in which the sum of the entropies of all bodies taking part in the process is increased."

Entropy is a scientific term describing the amount of disorder or randomness in a system, and measuring it provides insight into the direction of spontaneous change for many phenomena.

The 2nd law of thermodynamics states that energy in a system (like our universe) will always move from being concentrated in one place to spreading out evenly over time. This is also known as "increasing entropy" or "creating disorder." The direct consequences of this energy usage and the direction of change in natural systems can be summed up in two main analogies that are observable everywhere in our lives:

1.) ... "there is no way to fully (100%) convert any amount of energy into work"
2.) ... "heat cannot spontaneously flow from a colder material or region to a hotter material or region."

The two statements may not sound anything like

Planck's, but I can assure you that they reflect the same thing. The first one says that no matter what we do, there will always be energy losses, while the second states that energy tends to flow from hot to cold and never vice versa.

Both facts are well known to everyone from everyday life experience. The first statement can be tangibly experienced by placing your hand on the hood of a car following a lengthy winter journey. Despite the manufacturer's concerted efforts to optimize the vehicle's energy efficiency and fuel economy, you can still feel the warmth radiating from it—a testament to the energy loss from the internal combustion engine.

The second statement is why a cup of hot coffee will get colder the longer you leave it on the table, or an ice cube will always melt outside the fridge. The coffee will spread its energy to the atmospheric air, which is colder, while the ice cube will receive heat from the surrounding air, which is hotter.

But it's valid to ask, why can't the ice cube release the little heat that it still has when we take it outside the freezer and stay intact indefinitely? This doesn't violate any of the other accepted and proven laws. So, why does this never happen? This question may sound silly as everyone knows that heat only goes from hot to cold and never the other way around. However, this universally accepted fact is not inherently logical in isolation. What compels heat or any form of energy to transition from one object to another? Why don't all materials and substances retain their current state without releasing any energy? Wouldn't this stasis appear more logical, even more 'natural,' if you will?

<u>Deeper Explanation of the Law</u>

Before delving deeper, it's crucial to familiarize yourself with the concept of entropy. Entropy serves as a measure of disorder within a system. Let's illustrate this with a relatable analogy:

Consider a football match where two teams of eleven players occupy opposite sides of the field, separated by the central line. At this initial stage, entropy, or the level of disorder, is at its minimum. The 22 players are neatly arranged in two distinct groups, each on their designated side of the field.

Now, picture the referee blowing the whistle to commence the game. The players begin to move, intermingling and blending as they chase the ball. At this subsequent stage, we observe that the entropy on the football field has escalated.

Now, envision each player as a molecule and the two teams as separate gases within a box, with the central line acting as a separating membrane. When the match or 'time' starts, the membrane dissolves, allowing the molecules or 'players' of the two gases to mix within the box. This increases the system's disorder, amplifying entropy within this defined or 'closed' system.

Having grasped the concept of entropy, you can better comprehend the essence of the second law of thermodynamics. It asserts that all atoms and molecules within any material or object are naturally inclined toward a state of higher entropy. This indicates that everything in our surroundings is gradually or swiftly integrating with its environment, both in terms of energy and material composition.

Phenomena such as rusting, decaying, rotting, and aging are all visible demonstrations of the second law of thermodynamics. We describe these processes as occurring "spontaneously"—plainly, they transpire without our intervention or any external energy input.

To counteract this inherent spontaneous nature, we need to consciously inject our own energy in a controlled manner. This action can slow or halt the increase in entropy, serving as a buffer against the inevitable march toward disorder.

Regardless of our efforts to maintain the efficiency of our systems, energy will invariably disperse into the surrounding environment. This fundamental rule applies universally, from the functioning of our biological cells to the operation of our car engines.

Every process that transpires involves energy, and a portion of this energy inevitably dissipates into the ether. This seemingly "lost" energy isn't truly vanished; instead, it contributes to an unyielding cosmic inevitability—the gradual progression towards maximum entropy for the entire universe.

A Game of Chances

So, is it the 'entropy factor that induces objects to indiscriminately share their energy without an apparent reason? The answer is no. Entropy is merely a term we employ to describe the level of disorder within a closed system. The true catalyst for this ubiquitous blending is chance.

Put simply, the odds of a glass bottle dropping and shattering vastly outweigh the likelihood of sand molecules assembling under sufficient pressure and heat to 'spontaneously' form a conveniently shaped and sized glass bottle. The probability of encountering ordered states in our environment is significantly dwarfed by the likelihood of finding states of disorder. They are just too numerous!

As time progresses, all entities naturally gravitate towards disorder, escalating their entropy in accordance with the second law of thermodynamics.

<u>Everyday Examples of the 2nd Law of</u>

<u>Thermodynamics</u>

Here's a list of everyday life examples of the 2nd law of thermodynamics that will help you better understand how it affects us and defines our lives, often without realizing it.

• Evaporation of Liquids

Leave a bowl of water on a table, and you'll find that the water level has lowered after several days. This is not due to someone drinking from it, but rather a result of evaporation. The process increases the room's entropy as water molecules blend with the surrounding air molecules.

• The Refrigerator

Our refrigerator is an iconic representation of our attempt to curb the increase in entropy of the food stored within this 'closed' and regulated environment. We feed it electrical energy, which the compressor uses to circulate freon or other heat-exchanging fluids, extracting heat energy from the contained food and subsequently reducing the motion of their molecules. To clarify, the refrigerator itself is not a 'closed' system as it continually exchanges energy with its surroundings, decreasing entropy in its compartments while increasing entropy in the kitchen.

- The Rock Toss

Even an action as simple as throwing a rock contributes to increased entropy. By picking up the rock and tossing it, the chemical energy from your last meal is transformed into kinetic energy within your arm muscles, which then gets transferred to the rock. The rock loses some of its energy to air resistance, which marginally heats the air it travels through, and then to the ground upon impact, creating sound and transferring kinetic energy to nearby dust and gravel/sand particles. Consequently, the energy that the rock imparts to the ground is less than what your body expended to lift and throw it, with tiny amounts of energy lost to heat your muscles, warm the air, and agitate the ground.

- Sweating

Imagine sitting in a room filled with people. As the room heats up, your body starts sweating. The sweat absorbs heat from your body and evaporates, dispersing heat and sweat molecules into the room's air, thereby increasing its entropy.

- Rusting Nail

Leaving an iron nail exposed to the atmosphere results in the gradual formation of iron oxide on its surface, commonly known as 'rust.' This interaction between iron and atmospheric oxygen releases energy into the environment, signifying a spontaneous increase in the world's entropy through a typical chemical reaction.

- Consuming Food

Our food is converted into chemical energy, which powers our muscle movements. However, our bodies are somewhat inefficient 'machines,' converting only about 40% of the food's energy into movement and organ function. The remaining 60% dissipates into the atmosphere as heat and sweat, making us intrinsic contributors to the world's entropy increase.

- The Sun

Our sun and all stars across countless galaxies in this vast, predominantly cold universe are unique cases of concentrated energy and plasma (energized gas) radiating into space. This steadily increases the universe's entropy until we reach the maximum possible entropy value. This will occur when all stars have exhausted their fuel, with the last remaining one exploding as a supernova, dispersing material into space. After this event, the entropy of our entire universe will stabilize at its maximum limit.

The examples provided illustrate that the second law of thermodynamics underlies every minor and major process in our world. Everything adheres to the principle of perpetually escalating disorder, from the operations of our cellular systems to the dispersal of energy by our sun in the form of light and heat. We are powerless to halt this progression, and so we must learn to accept it. This inherent propensity towards chaos initially facilitated our existence, and paradoxically, the same principle guarantees our eventual demise.

Chapter Two – The Scientific Theory of Evolution

How The Theory of Evolution Came to Be

Although the conjecture that one living organism could evolve from a different kind has roots stretching back to the era of ancient Greek philosophers, Charles Darwin, the esteemed English naturalist, and geologist, first provided persuasive evidence for evolutionary theory. This evidence was presented to the world in his seminal work, "On the Origin of Species," published in 1859. Shortly after, in 1865, Austrian scientist Gregor Mendel delivered compelling proof, through meticulously conducted pea plant experiments, that hereditary traits are predictably passed on via biological elements, later known as genes. However, the time's limited scientific understanding and technological capabilities presented a significant challenge to accepting these revolutionary ideas. It wasn't until the 1930s that evolution, natural selection, and Mendelian inheritance were synergistically merged to form a comprehensive scientific biological framework, now renowned as the "Theory of Evolution."

<u>What Does the Theory of Evolution Say</u>

The "Theory of Evolution" proposes that all humans, animals, and plants originated from a common ancestral microbe or a similarly rudimentary organism that proliferated four billion years ago. Since this genesis, evolution has incrementally crafted an array of diverse species, with the majority becoming extinct and only those possessing advantageous traits and characteristics surviving the relentless struggle for existence.

The transition from biological simplicity to increasingly complex forms with each succeeding generation relies on the concept of DNA mutation during cell replication. These mutations are deemed random; hence, evolution continually augments the diversity among organisms, enabling the generation of myriad variations or subtly differentiated versions of the same entity. Depending on the environmental compatibility of a mutation, a species may persist and reproduce over centuries or face rapid extinction.

A quintessential example illustrating the mechanics of the "Theory of Evolution" is the transformation of wild wolves into domestic dogs. Genetic evidence suggests that dogs and modern gray wolves diverged from a now-extinct common ancestor, the Taimyr wolf, approximately 40,000 years ago. Human hunters subsequently selectively bred dogs, leading to the many distinct breeds we see today. Consequently, even the diminutive Chihuahua, despite its striking dissimilarity to the wild gray wolf, remains a close relative in biological terms and genetics.

Is the Theory of Evolution Valid?

The Theory of Evolution is widely accepted within the scientific community as the most comprehensive and consistent explanation that aligns with findings across various fields, such as biology, anthropology, genetics, paleontology, medicine, biochemistry, and molecular biology. It is necessary to clarify that a 'scientific theory' refers to a well-established understanding of a phenomenon backed by extensive evidence gathered through repeated testing, observations, and experiments.

Thus, the Theory of Evolution is not merely a speculative idea or an abstract concept; its facets are consistently and robustly confirmed by a global network of scientists. Unfortunately, the term 'theory' often leads to misconceptions, leading some to wrongly equate the Theory of Evolution to a mere hypothesis, overlooking its rigorous scientific foundations. This misunderstanding also extends to other well-established scientific theories like the Theory of Relativity, Thermodynamics Theory, Plate-Tectonics Theory, Quantum Field Theory, and the Atomic Theory.

Furthermore, it's no accident that the Big Bang Theory, which also rests on solid scientific ground, is sometimes dismissed by individuals with staunch religious convictions who misunderstand the term 'theory' in this context too.

While Darwin relied on his reasoning and observations, and Mendel sought to experimentally prove the existence of invisible factors that determine genetic traits, neither had access to the advanced tools that scientists have today to validate the Theory of Evolution. Techniques used in DNA analysis by biochemists and molecular biologists, such as restriction fragment length polymorphism and short tandem repeat analysis, have drastically revolutionized our ability to delve into DNA sequences, quantify tandem repeats, and draw reliable conclusions from the comparisons of different samples. These capabilities would have been inconceivable to people a century ago.

Furthermore, the domain of paleontology has also seen significant advancements. Paleontologists now use fossil samples of early life forms and employ radiocarbon dating techniques to estimate their origins. These tools first came into existence around the 1940s, nearly 80 years after Darwin published "On the Origin of Species." Using these methods, we now know that dinosaurs went extinct about 65 million years ago, and the oldest fossils discovered date back to approximately 230 million years ago.

The mounting evidence supporting the Theory of Evolution accumulates daily, emerging from scientific communities worldwide. These researchers are in an advantageous position, able to observe the rapid evolutionary traits of microorganisms that undergo changes over short time frames.

Unlike the need to wait centuries to observe minor physical changes in larger organisms, like, for example, as a millimeter length increase in an Arctic fox's ear tip indicating adaptation to changes in regional wind patterns, researchers working with microbes and invertebrates can witness significant developments occurring within just a few days, if not quicker.

Typically, to observe evolution in action, a relatively simple organism is required, alongside an environment to which this organism must adapt for survival and growth. Even though we may not all be scientists working in labs, we can still identify examples of organisms whose evolution has become noticeable within our own lifetimes.

Pesticides, a collection of herbicides, insecticides, and fungicides, have been globally employed for roughly eighty years to boost crop yields by eliminating organisms that threaten our produce.

The extensive use of these chemicals has led to noticeable genetic changes in the targeted pests. Many of these insects have developed a resilience to our chemicals, prompting scientists to create new seeds designed to withstand more potent pesticides, thus safeguarding our food supply. However, ensuring these stronger chemicals won't adversely affect our health remains challenging.

→ The seminal case of DDT, an organic insecticide introduced in 1945, epitomizes insect resistance development. Initially, DDT's efficacy was so unprecedented that scientists considered pest control a solved issue, ready to be consigned to history. However, only six years into the widespread application of DDT, farmers across Europe, Asia, Africa, and Central America began to observe a drastic reduction in its effectiveness. The common housefly emerged as one of the first species to adapt to DDT, with some cases reporting resistance as early as 1947, merely two years post its mass deployment. The most striking instance of DDT resistance, though, was found among malaria-carrying mosquitoes, a species capable of causing human fatalities.

➜ In England, an emblematic example of pesticide resistance involves a larger, more complex organism: the rat. These rats, colloquially termed "super rats," can ingest up to five times the amount of rat poison that their counterparts in other countries can consume, and they have also grown to a staggering length of 0.6 meters, justifying their imposing title. Initially, farmers attempted to counter this by increasing the quantities of rodenticides, but the rat population continued to burgeon, leading to dwindling crop yields year after year. Many farmers have shifted towards more eco-friendly solutions, such as deploying vigilant felines (perhaps, 'super cats'?) or simply diverting the rats' attention with a more enticing bait, thereby attempting to safeguard their crops.

But we need not delve deep into the animal kingdom to find examples of this phenomenon; Evidence of active evolutionary mechanisms abounds within our species, like:

- The Himalayan ethnic group of Sherpas has developed a larger-than-average respiratory system, allowing them to thrive in low-oxygen environments.

- The indigenous Bajau people in Indonesia have developed larger spleens, ideal for lengthy free-diving, fish-catching sessions.
- An additional artery that temporarily runs down the center of our forearms while we're still in the womb, which scientists at the University of Adelaide noticed appears in people today far more frequently (30%) compared to 1885 data (10%).

These instances serve as compelling evidence of human evolution. However, these traits demand substantial time to emerge and rigorous observation to confirm. Fortunately, we can witness evolutionary theory at work within our bodies in more readily observable ways.

Our well-being is inherently connected to the efficacy of antibiotics prescribed by healthcare professionals during our illnesses, and many of us have become familiar with the term "antibiotic resistance." Indeed, some have even witnessed its implications first-hand. Regrettably, organisms like bacteria, viruses, fungi, and parasites have demonstrated their ability to develop resistance against our pharmaceutical defenses.

A stark illustration of this reality is the HIV virus, first clinically observed about four decades ago. Despite advances in medical research, a cure remains elusive, primarily due to the virus's remarkable adaptability to antiviral treatments. Laboratory studies reveal that when faced with an antiviral drug, HIV rapidly mutates to produce drug-resistant strains. When challenged with multiple drugs, the virus counters with multi-drug resistant strains.

The genetic makeup of HIV consists of merely two strands of RNA, coding for the virus's nine genes, a representation of simplicity in genetic terms. This reality facilitates rapid evolution, making HIV a perfect case study for observing evolution in action at the smallest detectable scale, illustrating how quickly it can manifest from our human perspective.

Interplay Between Chances and Selection

In 1928, Dr. Sir Alexander Fleming discovered penicillin, marking the advent of humanity's use of antibiotics. However, he presciently warned of the emergence of penicillin-resistant microbes. Fleming foresaw that the widespread availability of penicillin might lead to improper usage, with many people self-administering sub-optimal doses, thereby providing an environment conducive to developing drug resistance in microbes.

This scenario was inevitable and unavoidable, much like an ice cube left out of the freezer is destined to melt. Regardless of size, complexity, or lack of a cognitive system, organisms tend to adhere to survival patterns. They adjust to their environment and conditions in an uncannily strategic way. While not all can adapt swiftly enough, some invariably do. From that point onward, the proliferation rate of the species becomes a decisive factor in cementing its robust presence in our world.

One might wonder, "Why don't these organisms just succumb?" How can bacteria, devoid of conscious will, develop resistance? Why does HIV persistently thwart our medical advancements, resulting in the loss of countless lives despite our relentless efforts? Is this just a roll of the cosmic dice, as we've previously explored in our discussion on the second law of thermodynamics? Or is there some inherent adaptive mechanism that springs into action as we strive to tackle these issues with our current technological arsenal?

→ Under the theory of evolution, DNA mutations within a species occur randomly. Therefore, the larger the population, the greater the likelihood of encountering multiple mutations, thereby expanding the genetic diversity and unintentionally fostering the development of resistance. For instance, the common housefly: capable of laying between 450 to 1200 eggs in a mere four days, the potential for genetic variation and mutation within a population of just one million houseflies in a single week is staggering. It's hardly surprising that they were among the pioneers in developing resistance to DDT.

→ The scenarios involving bacteria, viruses, and cancer cells reflect a similar theme. Bacteria, for instance, can exhibit significant genetic diversity due to their sheer volume within an organism. To illustrate, consider that a typical bacterium divides approximately every 20 minutes. This rapid multiplication gives an idea of the scale of a bacterial population within an organism during a few hours of illness. Since antibiotics usually target a specific enzyme within the bacteria, some bacteria will inevitably survive the treatment due to a mutation in the genetic code of this particular enzyme. Even if a small fraction of bacteria survive, they could continue to proliferate, this time immune to the same antibiotic. This illustrates why feeding livestock copious amounts of diverse antibiotics, as is done today, is counterproductive. The bacteria within them develop resistance and could potentially transfer to humans, leaving us with diminished treatment options.

Applied mathematical principles, such as probability theory and statistics, have proven valuable in predicting life's occurrence. For instance, the Poisson distribution, initially put forth by the French mathematician Denis Poisson in 1837. Today, microbiologists use this distribution to forecast microbial concentrations from quantal data. It enables them to precisely predict the volume of microbes contained in culture tubes, the probability of these samples being positive, and the dispersion of these organisms.

Poisson distribution is relevant in numerous biological scenarios, such as predicting the number of bacterial colonies on a Petri dish or the quantity of nucleotide base substitutions in a gene over a specific time frame. It also applies to more intricate organisms and larger systems, such as the number of trees in a given land area, the number of offspring a Geneva resident might have, or the incidence of congenital disabilities and genetic mutations at a clinic within a year.

Mathematical models like the Poisson distribution can help us understand and predict some aspects of biology, however, it is important to note that they can't capture all the complexities of biological systems, nor do they constitute explanatory frameworks.

While evolutionary processes are subject to chance, i.e. randomness in mutations, evolution is an interplay between random and decidedly non-random processes like natural selection. Organisms that carry beneficial mutations are more likely to survive and reproduce, thus passing on these advantageous traits to subsequent generations. Over time, this leads to the adaptation of populations to their environment, a process guided by the non-random force of natural selection.

A Driving Force?

Having a large population benefits all organisms by enhancing the odds in their favor. However, the rapid development of drug-resistant strains by HIV or the remarkable increase in active efflux by cancer cells, essentially pumping out anti-cancer drugs, is profoundly striking. These phenomena exemplify a kind of intrinsic programming that propels every living entity, even at the cellular level, towards survival to contend for existence until the very end.

The existence of this inherent trait and the mechanisms underlying it, its origin, and its evolution over time remain enigmatic. This mystery remains unsolved by concrete scientific evidence, leading many to regard it more as a facet of biological philosophy than empirical biology.

Intriguingly, recent studies propose that consciousness, and consequently an innate predisposition towards survival, may be embedded within all entities, encompassing even ostensibly lifeless matter. The philosophical standpoint of "panpsychism" is increasingly acknowledged by a growing number of esteemed neuroscientists, physicists, and philosophers.

The phenomenon of quantum entanglement serves as a potent illustration. It denotes a circumstance wherein two particles of a pair exhibit intertwined quantum states that collectively harmonize their behavior under a unified representation of their overall quantum state. In simpler terms, two "entangled" particles are interconnected in a manner that permits the exchange of information, facilitating the preservation of distinct states that ultimately uphold the conservation of mass, momentum, energy, and adherence to physical laws.

While all efforts to create analogies that fully capture all aspects of quantum entanglement are destined to fail due to the topic's notorious difficulty in illustrating with macroscopic examples, I will still make an effort with the below example:

→ Suppose you roll the dice inside two opaque cups and then separate the cups by a vast distance. Without lifting the cups, you don't know the result of either dice roll (similar to the uncertain quantum state of each particle). However, if you lift one cup, you instantly know the result of the other, as the total sum has been predetermined to be seven. For instance, if the first dice roll is a four, the other must be a three, no matter how far apart.

Distance has always been a focal point in experiments involving entangled particles. About ten years ago, physicists from the University of Geneva embarked on a groundbreaking experiment. They chose to observe the behavior of a pair of entangled photons separated by an expansive distance of 18 kilometers. The first photon embarked on a journey via optical fiber to the village of Satigny, whereas its pair made an identical voyage to the village of Jussy. The experiment's outcomes were astonishing.

The researchers observed that each photon seemingly "perceived" when its counterpart arrived at its destination. Furthermore, they discovered that the exchange of information between the pair transpired instantaneously, surpassing the speed of light. Before this, light speed was the upper limit for information transmission. This experiment, however, unveiled the existence of an inherent mechanism in all matter that transcends conventional boundaries and speed restrictions.

A 2017 experiment conducted by physicists at the

University of Sheffield proved that it's possible to place photosynthetic bacteria in a state of quantum coupling by using resonant light frequencies. This suggested that quantum theory doesn't apply only to small, subatomic particles but may also apply to larger systems, including biological organisms.

These scientific investigations provide evidence that mass possesses a form of consciousness, implying that all entities composed of mass are constructed from elements with a degree of consciousness. These fundamental building blocks lack intelligence, leading us to infer that consciousness does not originate from intelligence, nor does an intelligence threshold stimulate the emergence of consciousness.

If this holds true, consciousness must be directly tied to or be an immediate consequence of the enforcement and preservation of physical laws. Consequently, the instinct of all living beings to survive, regardless of their size and biological complexity, might be directly linked to the second law of thermodynamics. The existence of cells that consume and dissipate energy might appear inexplicable, but it is incontrovertibly connected to the fact that they increase the universe's entropy.

Though the above concepts are progressively

gaining scholarly recognition and acceptance, they remain contentious and largely reside within the sphere of belief, which leads us to the next chapter.

Chapter Three – The Religious Belief of Creationism

<u>Introduction</u>

The timeless and fundamental question, "How did we come into existence?" has long been the catalyst for fervent debates, impassioned arguments, and deep-rooted conflicts between creationists and evolutionists. Each camp diligently crafts and hones its arguments, drawing upon various scientific principles, with the theory of Thermodynamics assuming a significant role in these discussions.

In this chapter, we will delve into the assertions put forth by religious proponents, dissect the involvement of thermodynamics in our understanding of species' origins, and examine the validity of these claims. We'll explore what's substantiated and what aligns with logic, and finally, we will debunk any misconceptions that may arise.

<u>The Origins of Life</u>

The question of life's origin holds equal weight to its subsequent evolution. Creationists simplify this complex issue, attributing life's genesis to a divine architect—God—who they believe is responsible for creating the universe and every life form within it. Conversely, contemporary science posits two plausible theories for life's inception: the transformation of non-living matter into simple organic compounds, known as abiogenesis, and panspermia, the notion that life may have been 'seeded' on Earth through RNA-bearing comets or asteroids.

The abiogenesis hypothesis garners substantial support from the pioneering research conducted by Stanley Miller and Harold Urey in 1952. Through their innovative experimentation, they demonstrated how a cocktail of gases—akin to those thought to be present on primitive Earth, such as water, methane, ammonia, and hydrogen—could convert into organic compounds, the potential building blocks of life, under the influence of energy from lightning.

The panspermia theory also holds significant merit, bolstered by compelling empirical evidence. In 2022, Japanese scientists examining samples retrieved by the Hayabusa2 space probe identified 23 distinct types of amino acids—key components for life. These findings bolster the plausibility of Earth being 'contaminated' by life-bearing asteroids, thereby contributing to life's emergence on our planet.

The Numbers Today

The researchers we referenced in previous chapters of this book have made peace with witnessing the rapid evolution of bacteria and viruses within mere hours. Their work is firmly grounded in the Theory of Evolution, enabling them to innovate new pharmaceuticals, antibiotics, pesticides, and resilient seeds. Their daily operations are unhindered by doubts surrounding the tools they employ or the validity of evolutionary theory. Consequently, it's unsurprising that approximately 99.85% of the expansive scientific community in the US accepts the Theory of Evolution. This statistic reveals the level of acceptance within the scientific realm but doesn't necessarily reflect humanity's collective stance. One might assume that the views of this knowledgeable group concerning evolutionary theory's validity would carry significant weight, yet these opinions are often disregarded or conveniently overlooked.

Large-scale surveys conducted over the past decade indicate that public acceptance of the Theory of Evolution varies widely. In Northern European countries, acceptance peaks at around 80%, falling to 75% in Germany, 65% in Poland, 50% in Bulgaria, 40% in the USA, and a mere 25%

in Turkey.

A separate study based on religious affiliation reveals further discrepancies. Acceptance of the theory is highest among Buddhists (81%), Hindus (80%), and Jews (77%). In contrast, the theory's acceptance rate is significantly lower among Muslims (45%), Protestants (23%), Mormons (22%), and Jehovah's Witnesses, who exhibit the lowest level of acceptance at just 8%.

<u>The Scriptures Have Spoken</u>

The belief that all life forms on our planet—humans, animals, and plants alike—were fashioned by a divine entity originates from the Book of Genesis (2:4-2:24, "The Creation of Man and Woman"). This conviction, rooted deeply in religious teachings, is widely recognized across major religions, including Christianity, Islam—with corresponding creationist passages in the Qur'an—Hinduism, and Judaism. These religions largely propagate the "Young Earth Creationism" (YEC) concept, which professes that all existence was divinely crafted a few thousand years ago.

However, geological discoveries in the 19th century began to challenge this belief, presenting evidence that Earth's age far exceeded the few thousand years suggested by biblical texts. As a result, some turned to the theory known as "Gap Creationism." This perspective maintains the concept of the "seven days of creation" from Genesis but interprets each day as a distinct 24-hour period separated by extensive time gaps.

Another interpretation, "Progressive Creationism," proposed in the early 19th century, holds that God progressively brought forth new life forms over hundreds of millions of years.

Other creationist philosophies include the "Intelligent Design" theory, suggesting that while species can evolve independently, divine intervention at some point heightened biological complexity to a level that allowed for human existence and made Earth habitable for the biodiversity we observe today.

Lastly, the "Theistic Evolution" theory integrates religious belief with scientific understanding. It supports the Theory of Evolution as posited by biologists but adds the perspective that this evolutionary process is constantly set in motion and guided by a divine hand.

Francis Collins, an esteemed physician-geneticist and proponent of Theistic Evolution, asserts in his book "The Language of God" that selfless altruism, an attribute unique to humans, provides evidence for divine involvement in human evolution. Conversely, prominent figures in the scientific community, such as neuroscientist and philosopher Sam Harris, contend that Collins neglects to consider biological theories such as "kin selection" and "exaptation" in his analysis. Harris proposes that elements of human morality and selfless love may stem from foundational biological and psychological traits, which were themselves products of evolution.

The Creationists' Misconception

Many creationists claim that the Theory of Evolution fundamentally conflicts with the well-known Second Law of Thermodynamics. According to this law, within a closed system, all components naturally trend towards a higher entropy state characterized by increased molecular disorder, decay, and degradation over time. We witness this in our daily lives as complex and structured entities gradually break down into simpler, often unusable, forms.

Understandably, creationists extrapolate this principle to living organisms, arguing that it's implausible for a life form to evolve from a simpler to a more complex state. They propose instead that these entities must have been the result of divine creation. However, this line of thought is predicated on a misinterpretation and oversimplification of the Second Law of Thermodynamics.

A more nuanced understanding of this law reveals that it does not contradict the Theory of Evolution. Instead, these two scientific principles can be seen as mutually affirming each other's credibility, a concept we'll delve into in the next chapter.

Chapter Four – Thermodynamics and Evolution

<u>Introduction</u>

Creationists often reason that the escalating complexity of biological systems over time is irreconcilable with the Theory of Evolution. They argue that this complexity contradicts the Second Law of Thermodynamics, which postulates a prevailing increase in molecular disorder and energy dispersion. Such an argument, however, typically stems from a fundamental misunderstanding or selective interpretation of this law.

The Second Law of Thermodynamics describes an increase in entropy within a closed system. However, conceptualizing Earth as such a system is a fundamental error. The Sun, a colossal external energy source for all terrestrial life, lies beyond our planet. Earth's atmosphere does not impede sunlight, which enables life as we know it.

The Sun provides the necessary energy for plants to thrive and reproduce, nourishing herbivorous species and enabling their survival and reproduction. This celestial body also warms the oceans, the atmosphere, and the ground, supporting a diverse range of life forms. The Sun enables us to capture "free" energy and conserve our hard-won chemical energy storage. We owe our existence to the dispersion of photons from the Sun, which increases the universe's entropy.

Moreover, the Sun has played a pivotal role in shaping the evolutionary trajectory of life on Earth. An evident example of this is how our eyes have evolved to perceive light wavelengths between 400 to 700 nanometers, corresponding to the spectrum of colors we can discern. Notably, this specific range of wavelengths accounts for more than 80% of sunlight after it has traversed Earth's atmosphere. This fact is not a coincidence but compelling evidence of our adaptation to specific planetary and solar conditions over millions of years. Our evolution bears the indelible imprint of the Sun's influence.

From Simple to Complex

You might still be pondering, given that the Universe—being a closed system—tends towards maximum entropy, how can anything within it evolve from simplicity to complexity? Assessing the total entropy of a complex organism as a "closed system" is challenging and, in fact, erroneous. As previously explained, terrestrial life forms do not exist in a "closed system" as they receive energy from the Sun. However, if we view an organism as an assembly of small, "closed system" units, such as cells, and understand that cellular chemical processes individually increase entropy, we can infer that an organism with more cells will exhibit a greater increase in entropy. Therefore, the theory of evolution is not at odds with the second law of thermodynamics; instead, they mutually validate each other.

Nonetheless, this concept can be perplexing because we don't see this principle manifest in everyday inanimate objects. For instance, bits of metal don't spontaneously assemble into complex objects like spoons, axes, or even smartwatches. However, the entropy of a car, composed of numerous pieces of metals and other materials, is no greater than the total entropy of its individual components.

In a closed system, where entropy can increase, simple forms become more complex—even when they don't involve living organisms. Take the Sun, for example. It increases the Universe's entropy by dispersing energy and mass in the form of light, heat, and plasma through hydrogen fusion—a nuclear reaction involving a chain of collisions. This process starts with two pairs of hydrogen atoms combining to form two atoms of deuterium, which then collide with two more hydrogen atoms to create two helium-3 atoms. Those collide once more to eventually form two helium-4 atoms.

The key in this process is that hydrogen-1 comprises one proton and one electron, while helium-4 consists of two protons, two neutrons, and two electrons. This transformation vividly illustrates how a system can evolve from simplicity to complexity so long as entropy increases throughout the process. This is precisely what happens during the Sun's hydrogen fusion, which releases gamma rays, positrons, and neutrinos.

We Are Efficient Entropy Increase

The Sun accounts for 98% of our Solar System's total mass. The remaining 2% of mass and energy is destined to be used and depleted, satisfying the inherent requirement to increase entropy. Life forms residing on planets such as Earth, where hydrogen fusion is untenable, are instrumental in perpetuating the process of entropy increase. The evolution of these life forms from simplicity to complexity is pivotal in amplifying the rate and efficiency of this increase.

A more complex organism can expedite the entropy increase within the confines of our Universe. Compare an amoeba's energy intake and subsequent dispersion to that of a blue whale, for example. It becomes apparent that the latter increases the Universe's entropy far more effectively and swiftly. Consequently, this observation prompts a plethora of intriguing new inquiries. For example:

Why haven't blue whales evolved to be even larger to optimize the increase of entropy in our Universe?

Indeed, considering the abundance of space and krill in the oceans, it's logical to wonder why blue whales, currently peaking at around 30 meters in length, haven't evolved to be larger. Scientific research suggests that a 30-meter, 200-ton blue whale consumes about 480,000 calories worth of krill during each feeding dive. If a whale were to grow beyond this size, the energy requirements for bodily functions, swimming, and feeding would escalate exponentially. While an increase in size would amplify the whale's energy intake, the energy expenditures would concurrently surge. This is why blue whales have evolved to their current size, which represents the zenith of their potential efficiency as creatures. Scientists have calculated the apex of this efficiency to be around 33 meters, remarkably close to their current average length. This isn't coincidental but a clear illustration of how all life forms are finely tuned to comply with the 2nd law of thermodynamics, optimizing the increase in entropy within their systems.

Why haven't all animals evolved into larger species like blue whales and elephants?

Nature's primary objective is to maximize the size of each organism while always keeping energy efficiency as the pivotal criterion and deciding factor. To that end, even the tiniest insects seize smaller parcels of energy that larger creatures cannot utilize effectively. Our environment teems with hundreds, millions, even billions of creatures, each contributing to the escalation of our

Universe's entropy. There's a necessity for creatures of all sizes, from the gargantuan to the minuscule. This biodiversity ensures that no snippet of energy remains unexploited, or in other words, unmetabolized, leaving no opportunity for entropy increase untapped. For example, even the minute particles of food lodged between our teeth, which we as larger organisms cannot process and metabolize, are readily consumed and metabolized by bacteria and microbes inhabiting our oral cavity.

If efficiency is paramount, why do some species like horned beetles and deer carry seemingly inefficient ornaments like antlers, sometimes to their detriment?

At first glance, these ornamental appendages that some animals sport may seem detrimental to survival, casting doubt on the theory of evolution and its connection to the second law of thermodynamics. Even Darwin confessed to a colleague that the existence of the peacock's tail perplexed him as he struggled to explain its evolutionary rationale. However, it's crucial to acknowledge that animals featuring these embellishments typically fall into two sub-categories: those with prominent ornaments and those with smaller or no decorations.

The latter group exhibits a natural selection

trajectory emphasizing energy efficiency and conforming to scientific principles, thereby securing the species' survival and continued evolution. On the other hand, the ornamental group has developed these features predominantly through sexual selection, a specific mode of natural selection.

Sexual selection is largely driven by the aesthetic preferences of the female members of a species, which is why ornamental and decorative features are predominantly observed in males. Essentially, male birds have evolved to sport more vibrant plumage, despite this making them more conspicuous to predators, because female birds showed a preference for mates with brighter colors, leading to greater reproductive success for these more colorful males.

Does the existence of female preference contradict the second law of thermodynamics?

Far from contradicting it, female preference works in harmony with the second law of thermodynamics. Essentially, the capacity to sustain and exhibit additional ornamental attributes signals strength, robustness, abundance, and the capability to support these enhancements.

Research on sexual selection has revealed that

species maintain a delicate balance, adhering to ornamental and minimalistic evolutionary paths. Each of these routes still respects the second law of thermodynamics: the first pathway features species that can rapidly elevate entropy levels, while the second one optimizes energy consumption while still facilitating entropy increase.

Both paths serve the law's fundamental premise, each uniquely. The ornamental course steps up the rate of entropy increase, influenced by female preference, while the minimalistic route ensures species longevity and survival during unforeseen adversities or rapid environmental shifts. Hence, both approaches remain harmoniously compliant with the second law of thermodynamics.

Does this apply to humans too?

Humans, too, illustrate the intriguing dynamics of sexual selection. This is evident in our acquisition and maintenance of expensive assets like houses, cars, boats, and the like, which while not essential for survival, indeed often pose a liability. This behavior, regardless of its potential risk or complications, accelerates the entropy of our world due to the increased consumption of goods and energy it incites. Moreover, such ostentation often enhances one's prospects in sexual selection.

Though we haven't evolved horns or colorful tails, we've developed alternative means to exhibit our capacity to expedite the world's entropy more rapidly than others, a trait many find attractive. Our advanced intellect has furnished us with ways to escalate the entropy of our world that previous species, such as Neanderthals, could not have fathomed.

We drive cars, fly planes, explore other planets, host Olympic Games, wage world wars, and even use(d) nuclear bombs. Our consumption habits demand the extraction of an ever-increasing amount of materials. Our global population has reached unprecedented levels, and our activities inflict significant harm on our planet. In essence, we are increasing the entropy of our world at a rate far faster than ever before.

Is the necessity for highly intelligent creatures like humans within this system to consume energy and escalate entropy far beyond what animals can or care to accomplish?

This question ventures into philosophical territory. If plants and animals are functioning as efficiently as possible to increase the Universe's entropy, what is the need for humans who could obliterate everything in an instant? Perhaps the answer lies within that very query. The destruction of our planet, distressing as it may be, also contributes significantly to the rise of the Universe's entropy—a highly effective method indeed. Moreover, one could even postulate that the Second Law of Thermodynamics offers a scientific solution to Fermi's Paradox. It could be that this law dictates the evolutionary trajectory of all species to a point where they develop sexual reproduction, thereby unlocking the capacity to escalate their system's entropy far more rapidly, even through self-destructive means.

An alternative interpretation could be that intelligent beings are more likely to understand and exploit concentrated energy sources like petroleum. No other animal on Earth apart from humans would tap into this potential, let alone comprehend its value, right? The extraction and consumption of these energy reserves may have harmful implications as we currently practice it, yet it undeniably accelerates the Universe's entropy. The same principle applies to all types of fuels.

Lastly, highly intelligent beings might eventually transcend the confines of their home planets, exploring and colonizing other solar systems and even galaxies given sufficient time. This venture would further amplify the Universe's entropy through space travel and colonization. An uninhabited planet receiving solar radiation and reflecting it as still usable energy could see this energy utilized and dispersed into irrecoverable fragments upon colonization. Similarly, a planet with untapped organic energy resources only waits for a colonizer to harness them, thereby hastening the Universe's entropy increase. Humanity's current objective is to colonize other planets before we exhaust our own resources.

Why doesn't life exist on other planets? If conditions are too extreme for the existence of life, does this imply that the 2nd law of thermodynamics and the entropy increase do not apply universally?

While I cannot definitively speak on the life-bearing potential of other planets due to our limited exploration and understanding of them, I can highlight that life on Earth has demonstrated remarkable resilience and adaptability, thriving even in the most hostile environments. This assertion casts doubt on any assumptions that seemingly harsh conditions universally preclude the existence of life.

In 1990, the renowned astronomer and physicist Carl Sagan demonstrated through an experiment using the cameras of the departing Galileo spacecraft that even with our advanced technology, we might fail to detect signs of life on our own planet, let alone others. This raises the intriguing possibility that life may exist in places we least expect or in forms beyond our current recognition.

A compelling case study of life's tenacity on Earth is the GFAJ-1 organism, a bacteria residing in California's Mono Lake—an environment notoriously high in toxic arsenic levels. Conventionally, arsenic is lethal to organisms as it rapidly dissolves phosphorus, a vital DNA building block. Astonishingly, GFAJ-1 has adapted to thrive in this hostile setting by substituting phosphorus with arsenic in its DNA structure—a feat unmatched by any other known organism.

This remarkable adaptability of life reinforces the notion that entropy increase persists, even under conditions we deem entirely inhospitable. Additional studies have indicated that GFAJ-1 still prefers phosphorus if available, suggesting that organisms might universally prioritize certain 'efficient' substances when available, but resort to alternatives if necessary. By this logic, it is plausible that organisms on other planets might have evolved entirely distinct DNA structures, relying on unfamiliar substances yet remaining fully capable of encoding genetic information and contributing to the universal entropy increase.

<u>All Coming Together</u>

You should now be able to draw a clear link between the Theory of Evolution and the Second Law of Thermodynamics, understanding their compatibility both within the realms of biology and physics, as well as philosophically. This unique perspective, underpinned by the concept of energy-based action, elucidates how nature itself steers species evolution. This process considers energetic efficiency to maximize the dispersion of usable energy units while ensuring the survival and evolution of species to the greatest attainable extent.

For those who ascribe to creationist views or hold deeply rooted religious beliefs, this work should not be perceived as a direct challenge to their faith. Instead, it should serve as a catalyst for further introspection, contemplation and as the starting point for enlightening discourse. None of the assertions made, or data presented within this book negate the possibility of a divine entity or universal creator. If such a God exists and was responsible for creation, then perhaps one of his most profound creations was the Second Law of Thermodynamics itself.